Matter Is Everything

Written by Becky Gold

Table of Contents

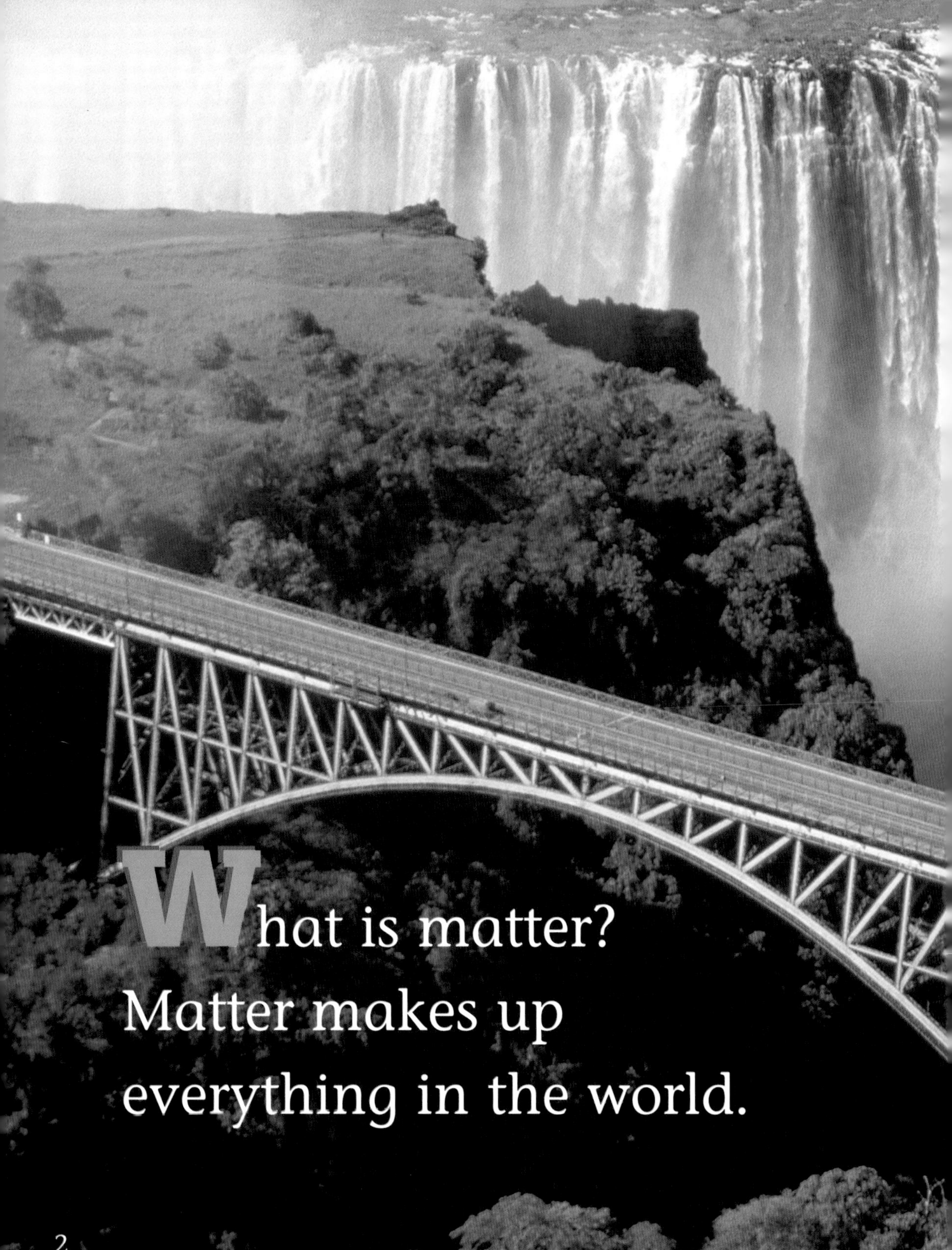

What is matter? Matter makes up everything in the world.

Matter is everything you can see or feel, taste or smell. There are three kinds of matter: solid, liquid, and gas.

A tractor is a solid.

Water is a liquid.

Air is a gas.

Solids

This picture shows a lighthouse on a rocky shore. The large boulders are solids. The lighthouse and its glass windows are solids, too.

Solid matter has a shape that you can see and feel.

What other solids do you see in these pictures?

Feather
Sneakers
Pepper

This big truck and this winter hat are both solids. But they don't look or feel the same. One is big and the other is small. One is hard and the other is soft.

Clay is a solid that is **flexible**. That means it is easy to bend and shape.

Some solids are powdery, like cocoa. Some solids are squishy, like a wet sponge.

Clay
Cocoa
Sponges

Liquids

How is a waterfall like syrup? Both are liquids. Milk, paint, and honey are liquids, too.

Liquids flow and take the shape of their container. They don't have a shape of their own.

Not all liquids look and feel the same. Some are thick, and some are thin. Some liquids flow more slowly than others.

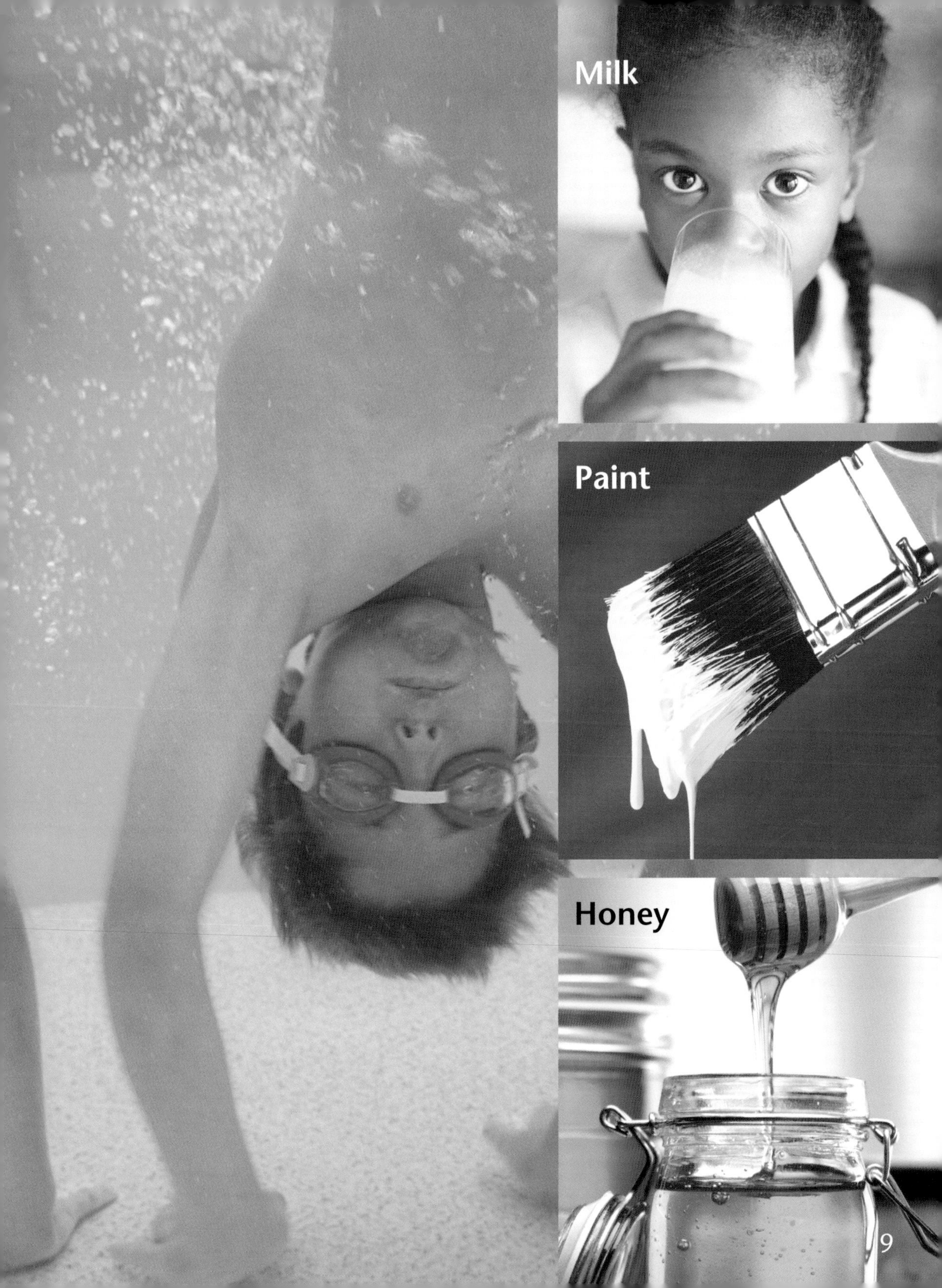
Milk
Paint
Honey

Gases

What would you call the air we breathe? It's not a solid or a liquid. You can't hold it, but you can feel it when the wind blows.

Air is a **gas.** Gases are often invisible. Like liquids, they take the shape of their container.

Air is a gas.

Steam is a gas.

Vapor is a gas.

Changing Matter

Now you know that matter can be a solid, a liquid, or a gas. But did you know that when matter gets very hot or very cold, it can change from one form to another?

Cold can change liquids into solids. What happens to water when it gets very, very cold? It freezes and turns into a solid, called ice.

Icicles form when dripping water freezes into long, thin sticks of ice.

This frozen treat started as a liquid. Cold turned it into a solid.

Heat can change solids into liquids. When ice melts, it turns into water. Even strong metals, such as iron and gold, can be turned into liquids by heating them at very high temperatures.

Heat can also change liquids into gases. Have you ever seen steam rising from a pot of boiling water? That steam is water that has turned into a gas.

Liquid gold

A geyser is hot water and steam that gush out of the ground.

Lava is hot liquid rock that pours out of a volcano.

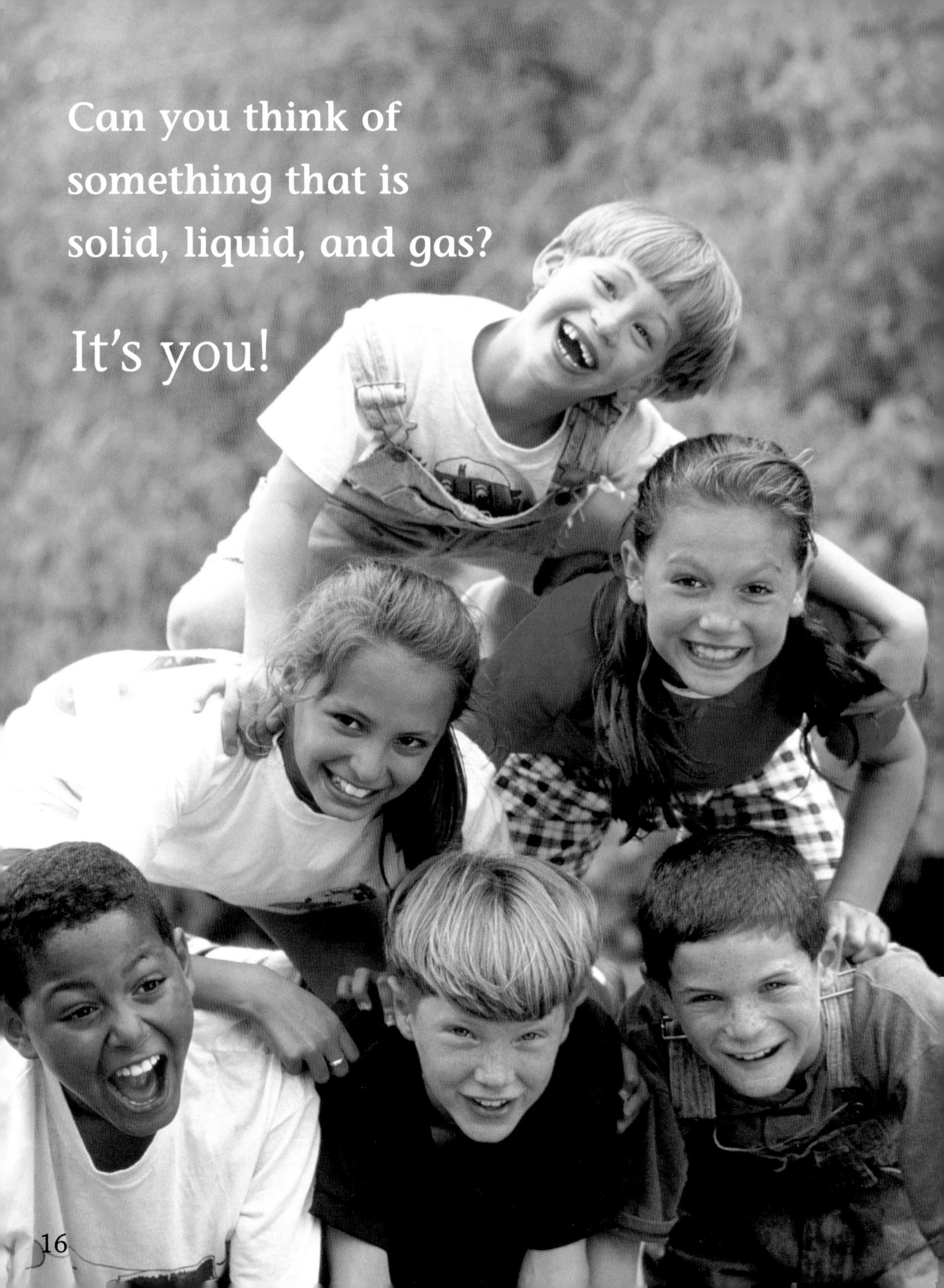

Can you think of
something that is
solid, liquid, and gas?

It's you!